YOUR KNOWLEDGE HAS VALUE

Maryna Psol

Tetraspanin2 is a candidate for compensation of PLP functions

GRIN Verlag

Bibliografische Information der Deutschen Nationalbibliothek:

Die Deutsche Bibliothek verzeichnet diese Publikation in der Deutschen National-
bibliografie; detaillierte bibliografische Daten sind im Internet über http://dnb.d-
nb.de/ abrufbar.

Dieses Werk sowie alle darin enthaltenen einzelnen Beiträge und Abbildungen
sind urheberrechtlich geschützt. Jede Verwertung, die nicht ausdrücklich vom
Urheberrechtsschutz zugelassen ist, bedarf der vorherigen Zustimmung des Verla-
ges. Das gilt insbesondere für Vervielfältigungen, Bearbeitungen, Übersetzungen,
Mikroverfilmungen, Auswertungen durch Datenbanken und für die Einspeicherung
und Verarbeitung in elektronische Systeme. Alle Rechte, auch die des auszugsweisen
Nachdrucks, der fotomechanischen Wiedergabe (einschließlich Mikrokopie) sowie
der Auswertung durch Datenbanken oder ähnliche Einrichtungen, vorbehalten.

Imprint:

Copyright © 2012 GRIN Verlag GmbH
Druck und Bindung: Books on Demand GmbH, Norderstedt Germany
ISBN: 978-3-656-51383-4

This book at GRIN:

http://www.grin.com/en/e-book/233026/tetraspanin2-is-a-candidate-for-compensa-
tion-of-plp-functions

GRIN - Your knowledge has value

Der GRIN Verlag publiziert seit 1998 wissenschaftliche Arbeiten von Studenten, Hochschullehrern und anderen Akademikern als eBook und gedrucktes Buch. Die Verlagswebsite www.grin.com ist die ideale Plattform zur Veröffentlichung von Hausarbeiten, Abschlussarbeiten, wissenschaftlichen Aufsätzen, Dissertationen und Fachbüchern.

Tetraspanin 2 is a candidate for compensation of PLP functions

Protocol of laboratory rotation

Submitted by Maryna Psol

2012

ABSTRACT

Tetraspanin2 (Tspan2) is a member of the tetraspan/transmembrane4 superfamily restricted to the nervous system. The abundance of Tspan2 is low in physiological condition but increases greatly in myelin of PLP-deficient mice which may indicate its' compensatory functions. Our experiments show that Tspan2 has no effect on delay of myelination at P14; young and old mice lacking Tspan2 has the same g-ratio as wild type. However, the quantity of unhealthy axons and axonal swellings are higher in aged animals lacking both PLP and Tspan2 than in single PLP knockout. Furthermore, the absence of PLP causes the decrease in the number of axons and rise in axon diameter; lack of Tspan2 alone leads to decrease in quantity of 0.4-0.7 µm diameter axons in 40 weeks old mice. These findings point to a possible auxiliary role of Tspan2 in support of axonal transport and long-term axon preservation in PLP^{null} condition.

INTRODUCTION

Myelination is the process of formation of multilayered myelin sheaths around the axons, mediated by oligodendrocytes in the CNS and Schwann cells in the PNS. Well known is the function of myelin in electrical insulation for rapid propagation of action potentials along nerve fibers. Moreover, myelin ensheathment is thought to be necessary for normal axonal transport, functional integrity and long-term survival of axons. Demyelination results in characteristic impairment of motor and sensory functions and lead to severe neurological disorders, e.g. leukodystrophies, neuropathies, and multiple sclerosis. Also subtle changes in myelin structure may contribute to psychiatric disorders including schizophrenia (Nave, 2010).

Being a highly specialized structure, multilayered myelin sheath requires coordinated integration of large quantities of specific proteins and lipids (Bosse et al., 2003). The most abundant proteins of CNS myelin are proteolipid protein (PLP) and its smaller isoform DM20 (Yool et al., 2001). According to the latest data PLP constitutes 15, 43% of the total myelin protein (Werner et al., in press). Despite of its abundance, the biological role of PLP is not completely understood. It was shown that PLP^{null} mice synthesize fully functional myelin with only minor ultrastructural changes of the intraperiod line and lower stability (Klugmann et al., 1997). Also in the absence PLP/DM20 small diameter axons are prone to axonal swellings and degenerations, which indicates role of mentioned proteins in the maintenance of axon-oligodendrocyte interactions. Some proportion of small diameter fibres doesn't form compacted myelin or myelinate with delay (Yool et al., 2001). On the other hand, overexpression of PLP induces severe neuropathology, including dysmyelination, progressive tremors, ataxia and premature death (Möbius et al., 2009).

Myelin membranes are unusually enriched in lipids (>70% of dry weight), and especially in cholesterol (>25% of total lipid content). Dynamics of lipid 'rafts' or cholesterol-rich membrane-microdomains modulate biological function of associated proteins and might have a crucial role in myelin biogenesis (Werner et al., in press; Gielen et al., 2006). PLP induces cholesterol accumulation by molecular association and co-transport, which, accordingly, facilitates myelination. However, the ability of PLP^{null} mice to produce fully functional myelin indicates existence of other proteins which compensate for PLP deficiency. A homolog of PLP, M6B, was shown to contribute to myelination and cholesterol enrichment in absence of proteolipid protein (Werner et al., in press). Moreover, mice lacking both PLP and M6B have limited myelination suggesting that other proteins may also have compensatory qualities. Among possible candidates are other transmembrane tetraspans, e.g. plasmolipin, CD9, CD81, CD82, tetraspanin 2 (Tspan2) (Werner et al., in press). Differential myelin proteome analysis revealed over 160 proteins in myelin, but only quantity of Tspan2 is greatly increased in myelin of PLP^{null} animals (Werner et al., 2007). This indicates that Tspan2 may be involved in myelination and neuroprotection in the absence of PLP.

The tetraspanin family includes proteins which have four hydrophobic transmembrane domains and two extracellular domains (Birling et al., 1999). They are found in all cells except erythrocytes and involved in cell differentiation, proliferation and motility, mediation of signal transduction, and, possibly, organization of membrane topology. Tetraspan proteins interact with each other forming so called tetraspanin webs at the cell surface. Precipitation with digitonin showed that these supramolecular complexes are also tightly associated with cholesterol (Charrin et al., 2003). Tspan2 is a member of tetraspanin family which is restricted to the nervous system and predominantly expressed in oligodendrocytes. It is detectable in the rats' brain from the postnatal day 3 and strongly increases in concentration to P22. High level of expression of Tspan2 at early stages of development and involvement in integrin signalling may indicate its role in differentiation of oligodendrocytes and myelin formation. Persistence of Tspan2 at later stages suggests that it may contribute to the stabilization of mature myelin, trophic support of axons and neuroprotection (Birling et al., 1999).

In current project we employed morphometric analysis of optic nerves cross sections of P14 and 40 weeks old mice in order to find out whether Tspan2 can reduce delay of myelination or axonal degeneration in mice with PLP-deficient myelin.

MATERIALS AND METHODS

Animals

Wild type, Tspan2^{null}, PLP^{null} and Tspan2^{null}PLP^{null} mice were included in the experiments. Constitutive mutants were bred into the C75BL/6 background using the mice from the breeding colony of the Max Plank Institute of Experimental Medicine. Only male mice of ages P14 and 40 weeks were used. Animals were sacrificed by cervical dislocation. All experiments were carried out in accordance with the animal protection law and approved by the German Federal State of Niedersachsen.

Electron and light microscopy

Optic nerves of 40 weeks old mice were fixed with 4% formaldehyde and 2% glutaraldehyde in PBS, and then postfixed with 2% osmium tetroxide. Samples were dehydrated in series of ethanol solutions of increasing concentrations. Ethanol was substituted with propylenoxid, and tissues were processed into epoxy resin according to standard procedures. Ultrathin sections (~50 nm) were cut on Leica Ultracut S ultramicrotome, stained with 4% aqueous uranyl acetate and lead citrate, and examined in Zeiss EM 900 electron microscope. 10 pictures per animal were collected at nominal magnification of 7, 000 with an on-axis 2048x2048 CCD camera. Pictures of P14 mice optic nerves were provided by Dr. Wiebke Möbius (5 picture per animal; LEO EM 912AB electron microscope). Embedded in epoxy resin optic nerves were also used to prepare semi thin (~500 nm) sections, which were stained with methylenblue-azur2 solutions and observed in bright field microscope.

Morphometric analysis of electron micrographs

Electron microphotographs were analyzed with the ImageJ (Fiji) software. In order to randomize axons we placed a grid on the picture and counted only axons crossed by the grating. For validation of axon pathologies all axons on the microphotographs were calculated.

Quantification of myelinated vs. nonmyelinated axons

To reveal a delay in myelination, we compared the proportion of myelinated and nonmyelinated axons in P14 mice. An axon was counted as myelinated if it was encircled at least by one compacted myelin layer. Axons were distinguished from other cells by their shape and characteristic appearance of cytoplasm (Figure 1). 700-1200 axons were validated per animal (3-4 animals in each group). The quantity of nonmyelinated axons was calculated as percentage of all validated axons.

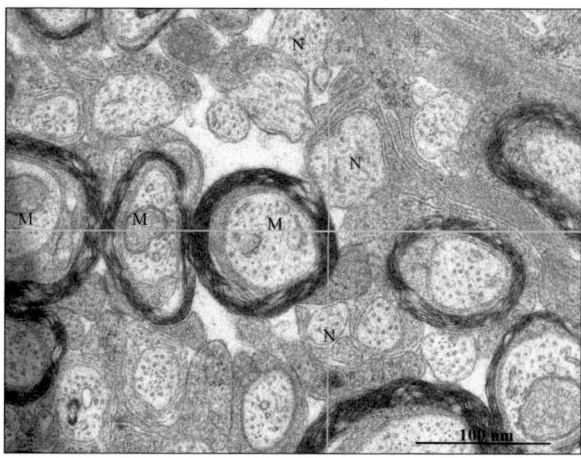

Figure 1. Transmission electron microscopy of optic nerve. Morphology of myelinated and nonmyelinated fibers. Axons appear as nearly circular profile in cross section. Microtubules, mitochondria, and elements of endoplasmic reticulum could be observed in axoplasm of myelinated (M) and nonmyelinated (N) axons. Processes of oligodendrocytes typically look darker than nonmyelinated axons. Only fibers crossed by the grid (green line) were calculated.

Validation of axon pathologies

We calculated the percentage of pathological axons in 40 weeks old mice in order to reveal if absence of Tspan-2 has effect on axon degeneration. Fibers were validated in 6 categories: healthy myelinated axons, healthy nonmyelinated axons, axons with increased adaxonal space, axonal swellings, unhealthy looking axons, and degenerating axons (Figure 2). Groups of axons with increased adaxonal space included fibers which had multivesicular bodies and other inclusions in adaxonal (periaxonal) space and/or between innermost myelin layers. Degenerating axons were identified by tubovesicular structures and amorphous cytoplasm. Axons which were not included in any other category (e.g. invaginations of the axons by the inner tongue of the myelin sheath) and those with mild signs of pathologies made up the group of unhealthy looking axons. Each fiber was assigned only to one category.

g-ratio and axon diameter

The g-ratio is defined as the ratio of axon Feret diameter to the myelinated fiber Feret diameter. Measurements were made from electron microphotographs of randomly selected fields of optic nerves of wild type and Tspan2^{null} mice of ages P14 and 40 weeks. A minimum 150 axons was assessed for each animal. Measurements were stratified by axon diameter in ranges of 0,4-0,7; 0,7-1,0; 1,0-1,3; 1,3-1,6; 1,6-1,9; 1,9-2,2; >2,2 μm. When using these ranges, the lowest values were always included and the highest ones excluded (e.g., the 0,4-0,7 range includes 0,4 through 0,6999, excluding 0,7).

The axon diameter was calculated for 40 weeks old mice of wild type, Tspan2^{null}, PLP^{null} and Tspan2^{null}PLP^{null} groups. A minimum of 150 axons was counted per animal and at least 4 animals were used per condition. Swollen or degenerating fibers were excluded from the measurements.

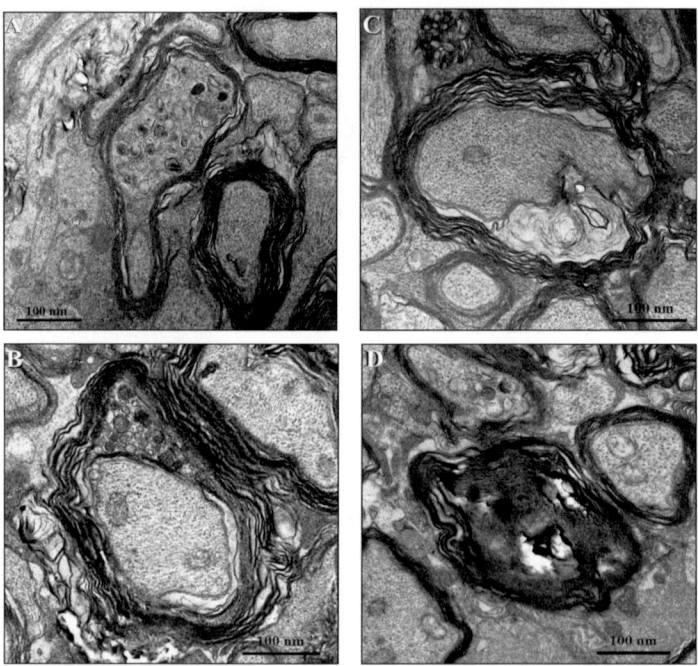

Figure 2. Axon pathologies in the cross sections of optic nerves

A – axonal swelling characterized by presence of numerous membraneous organelles and increased diameter; B – axon with vesicles in adaxonal space and swelling between innermost myelin layers; C – unhealthy looking axon; D – degenerating axon demonstrates loss of microtubules and appears darker on microphotographs.

Statistical analysis

GraphPad Prism software 5.0 was used for data analysis. Quantifications of myelinated and nonmyelinated fibers, axon pathologies and axon diameter were evaluated with one-way ANOVA with Bonferroni correction and two-tailed unpaired t-test with Welch's correction. Comparisons of g-ratio between genetic groups and axons of different diameter were performed using two-way ANOVA with Bonferroni post hoc test. The level of significance was set at $P \leq 0.05$ (*), $P \leq 0.01$ (**), and $P \leq 0.001$ (***).

RESULTS

Tspan-2 is not essential for myelination

Axonal assessment in P14 optic nerve revealed no difference in the percentage of nonmyelinated fibers between wild type and Tspan2^{null}, and between PLP^{null} and Tspan2^{null}PLP^{null} mice. The difference between wild type and PLP^{null} (p = 0,07), as well as between wild type and double knockout animals (p = 0,06) is not significant, but it is close to the statistical significance. Thus, we can say that there is a trend to have more nonmyelinated axons in PLP^{null} and Tspan2^{null}PLP^{null} mice. Yool et al. showed earlier that in the absence of PLP quantity of nonmyelinated axons is increased at P20, P60, and P120. The calculated percentage of nonmyelinated fibers at P14 amounts to 55-65%, which is consistent with previously published data (Yool et al., 2001). Overall, we can conclude that Tspan-2 provides no additional effects on delay of myelination.

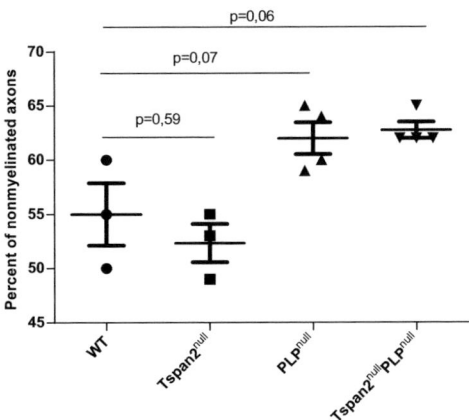

Figure 3. Percentage of nonmyelinated axons in P14 optic nerve

g-ratio does not differ between wild type and Tspan2^{null} mice

In order to evaluate the level of myelination in Tspan2^{null} mice we calculated the g-ratio, the ratio of axon diameter to myelinated fiber diameter. Figure 4 B, D shows that data points of g-ratio majorly overlap for two genetic groups with exception of few bigger axons in 40 weeks old Tspan2^{null} mice. g-ratio lies within the range of 0.6 - 0.94 in P14 and 0.57 – 0.9 in aging animals. Figure 4 A, B demonstrates that g-ratio of Tspan2-deficient mice is not significantly different from WT in any axon diameter stratum but the g-ratio slightly increases with growing axon diameter. Overall, axons of wild type and Tspan2^{null} mice were with myelin of equal thickness.

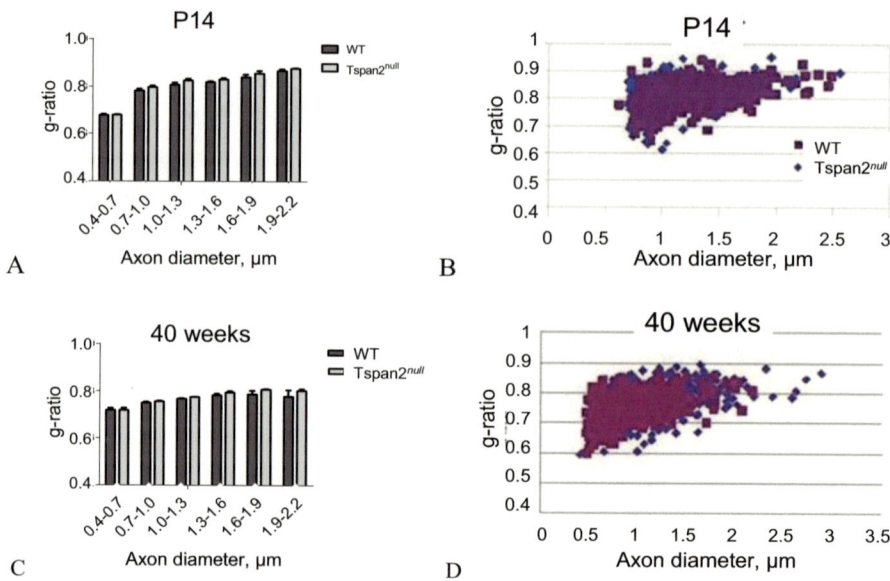

Figure 4. g-ratio in optic nerves of WT and Tspan2^{null} mice.

Axonal pathologies in aging mice

We have evaluated optic nerves of 40 weeks old mice on presence of pathologies to see if Tspan2 protects axons against degeneration. Used categories are described above in "Materials and methods". Figure 5 shows that single Tspan2^{null} animals are not significantly different from WT in all categories. Mice lacking PLP have less healthy myelinated and more unmyelinated axons compared to WT. Also absence of Tspan2 in PLP^{null} has no additional effect on level of myelination in aging animals. PLP deficient mice have larger than in WT percentage of axons with increased adaxonal space, but there were no statistically significant difference between PLP^{null} and Tspan2^{null}PLP^{null}.

Axonal swellings are virtually absent in WT and Tspan2^{null} mice but their amount increased up to 0.9 % in PLP and 1.2% in double knock out animals, which is significantly higher. Difference between PLP^{null} and Tspan2^{null}PLP^{null} is close to statistical significance (p=0,072) which might indicate a trend of having more axonal swellings in Tspan2^{null}PLP^{null} mice comparatively to PLP^{null}.

Percentages of unhealthy axons are significantly higher in Tspan2^{null}PLP^{null} in comparison with PLP^{null} mice (p = 0.001).The difference between these two groups and wild type is also significant.

Numbers of degenerated axons vary within the range of 1-6% with no significant difference between genetic groups. This could be explained with a fast phagocytosis of degenerated processes by microglia.

Taken together, increase of axonal swellings and unhealthy axons in double knockout mice suggest that Tspan2 may contribute to maintenance of axonal transport and thereby reduce axonal degeneration in the absence of PLP.

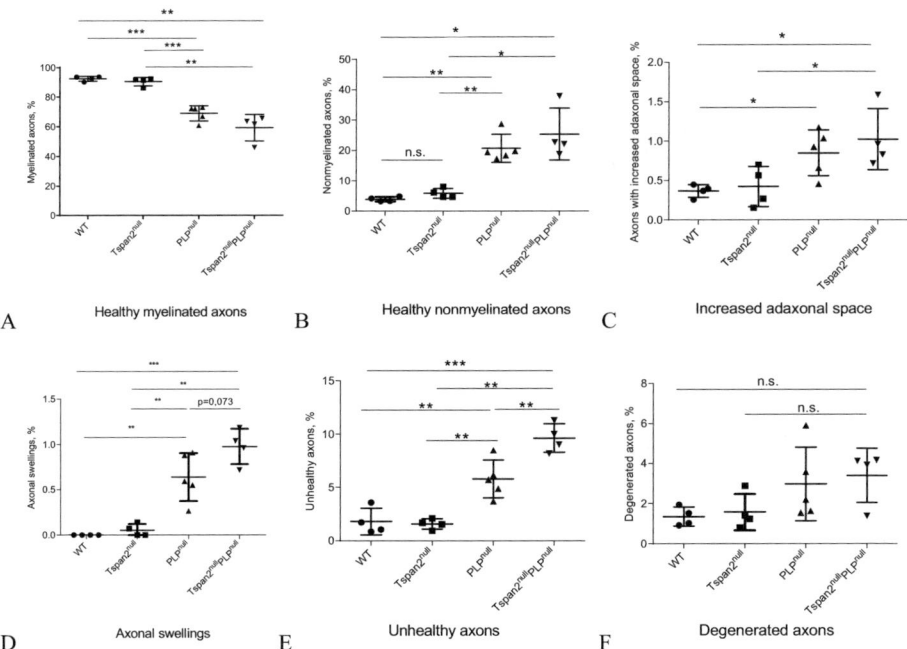

A   Healthy myelinated axons   B   Healthy nonmyelinated axons   C   Increased adaxonal space

D   Axonal swellings   E   Unhealthy axons   F   Degenerated axons

Figure 5. Axons with pathologies in 40 weeks old mice.
Diagrams demonstrate percentages of (A) healthy myelinated axons, (B) healthy nonmyelinated axons, (C) fibers with increased adaxonal space, (D) axonal swellings. (E) unhealthy, and (F) degenerated axons in WT, Tspan2^{null}, PLP^{null} and Tspan2^{null}PLP^{null}.
Note: P $\leq$ 0.05 (*), P $\leq$ 0.01 (**), and P $\leq$ 0.001 (***).

Quantity of axons decrease in PLP^{null} and Tspan2^{null}PLP^{null} mice

Estimation of axon number in optic nerve cross sections of 40 weeks old animals revealed a significant reduction of fibers quantity in mice lacking PLP. Absence of tetraspanin 2 provided no additional impact (Figure 6).

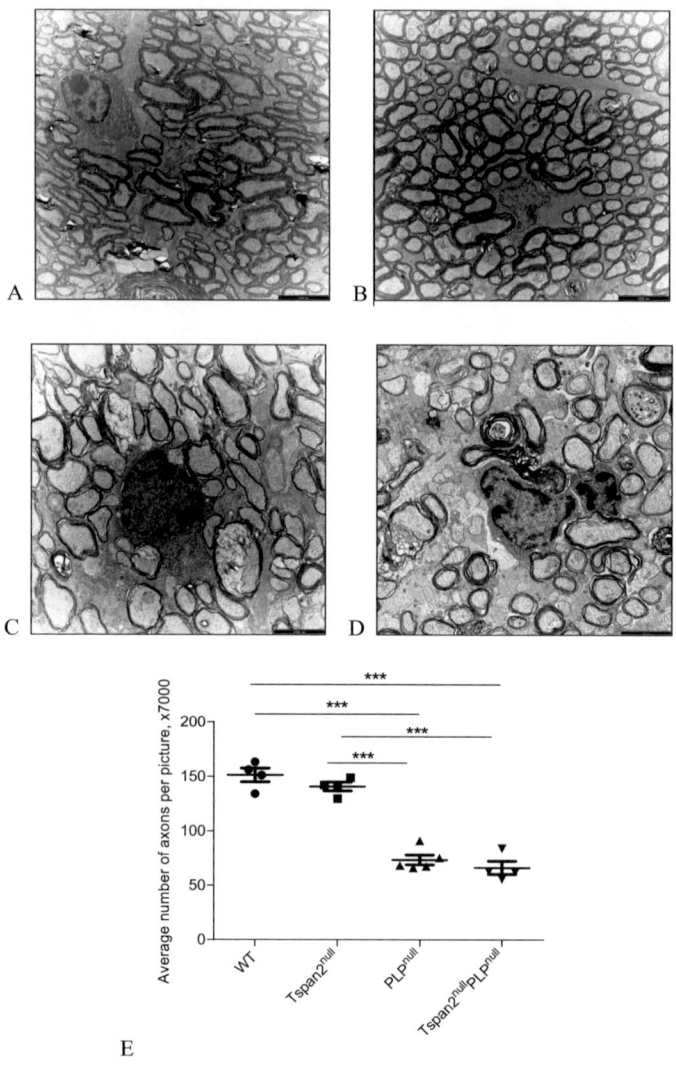

E

Figure 6. Number of axons in 40 weeks mice optic nerve cross sections.

Axons appear smaller and more compactly located in wild type (A) and Tspan2^{null} (B) mice than in PLP^{null} (C) and Tspan2^{null}PLP^{null} (D) animals. Scale bars = 2.5 μm

Fig. 4E demonstrates average number of axons per microphotograph in four groups of animals. The optic nerves of PLP^{null} and Tspan2^{null}PLP^{null} mice contain significantly fewer axons than WT (p=0,0002, p=0,0002) and Tspan2^{null} (p<0,0001, p=0,0002). There are no significant differences between WT and Tspan2^{null}, PLP^{null} and Tspan2^{null}PLP^{null}.

Note: P $\leq$ 0.05 (*), P $\leq$ 0.01 (**), and P $\leq$ 0.001 (***).

Axon diameter increase in absence of PLP

Analyzing the optic nerve microphotographs of aging mice, we noticed that axon diameters in PLP*null* and Tspan2*null*PLP*null* animals are higher than in wild type (Figure 6 A-D). Analysis of frequency distribution of axonal diameter with two-way ANOVA and Bonferroni post-hoc test revealed more details: PLP*null* and Tspan2*null*PLP*null* have significantly fewer axons with diameter 0.4-1.0 µm and more fibers of size 1.3-2.2 µm, Tspan2*null* mice have fewer axons with diameter 0.4-0.7 µm ($p < 0.0001$) in comparison to wild type, and PLP*null* and Tspan2*null*PLP*null* are not significantly different from each other.

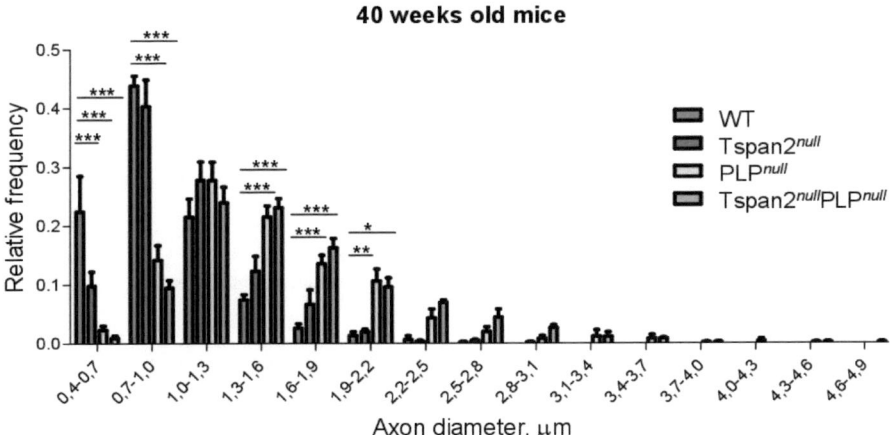

Figure 7. Axon diameter frequency distribution in optic nerve of wild type, Tspan2*null*, PLP*null* and Tspan2*null*PLP*null* aging mice. Note: $P \leq 0.05$ (*), $P \leq 0.01$ (**), and $P \leq 0.001$ (***).

DISCUSSION

Mutations of the *PLP1* gene are associated with the leukodystrophies, Pelizeaus-Merzbacher disease and Spastic Paraplegia 2. Besides PLP, secondary molecular alterations may facilitate or delay pathophysiology of these diseases (Werner et al., 2007). In this project we performed morphometric analysis of optic nerves cross sections to see if Tspan2 contribute to myelination and protection against axon degeneration in PLP-deficient mice. This is based on the previous observation that the abundance of Tspan2 is strongly increased in PLP-deficient myelin.

Our experiments showed that the proportion of nonmyelinated axons were the same in wild type and Tspan2^{null} mice as well as in PLP^{null} and Tspan2^{null}PLP^{null}. Although we noticed only a trend for delay in myelination in PLP-deficient mice, it was previously shown that in the absence of PLP a subset of small diameter axons fail to myelinate properly and, consequently, the quantity of unmyelinated fibers increase (Yool et al., 2001). We attribute such difference in the results to the variance in wild type data. Our P14 wild type mice were not littermates, and probably were different for several hours in age. This could contribute to the level of myelination. Nonetheless, Tspan2 clearly showed no effect on the delay of myelination in P14 mice.

Myelin sheath thicknesses were the same in optic nerve fibers obtained from Tspan2^{null} and wild type mice. According to Chomiak et al., the optimal g-ratio for the conduction in rat CNS is close to 0.77; our data is consistent with this value. Also fibers of different diameter had relatively stable g-ratio both in young and aged mice. Therefore, Tspan2-deficient mice demonstrate robust myelin assembly.

Viability and functional integrity of myelinated axons depend on extrinsic glial signals (Bjartmar et al., 1999). In our experiments the lack of PLP in aged mice promoted increased diameter of fibers, axonal pathologies and loss - the characteristics of disordered interactions between axons and oligodendrocytes. The mechanisms for specification of axon caliber are not well understood, but they depend on myelin thickness, number and composition of neurofilaments (Barry et al., 2012, de Waegh et al., 1992). Also radial axonal growth and accumulation of neurofilaments are constricted to the axonal regions enveloped with oligodendrocytes processes (Sanchez et al., 1996). In regard to histopathological changes PLP-deficient mice had more axonal swellings, fibers with increased adaxonal space and other unhealthy looking axons in the optic nerves. Griffiths et al. showed that PLP^{null} animals develop axonal swellings containing membraneous organelles, neurofilaments and multivesicular bodies after 6-8 weeks of age and predominantly in paranodal regions (Griffiths et al., 1998). Such accumulations impair slow anterograde and retrograde transport and eventually these pathological changes lead to the degeneration and progressive axonal loss on the second year of life (Bjartmar et al., 1999, Edgar et al., 2004). In fact we saw the dramatic decrease (~50%) in quantity of axons already at the age of 40 weeks in PLPnull mice comparatively to wild type.

Tetraspanin 2 only slightly affects the state of axons in PLP-deficient condition. Mice which lack both proteins had significantly more unhealthy looking axons and tendency for increased number of axonal swellings. This indicates a possible function of Tspan2 in support of normal axonal transport and oligodendrocyte-axon interactions. Tspan2 did not influence considerably the overall quantities of axons and axon diameter. Relatively low impact of Tspan2 could be explained by compensation with other tetraspanin proteins present in CNS myelin: CD9, CD81, CD82 and so on. Also, we still suppose that PLP^{null} mice maintain viable axons longer than double knock out animals due to compensatory role of Tspan2. But both of these genetic groups demonstrate rather severe morphological alterations at the age of 40 weeks. In order to see in which group degeneration develops earlier one should investigate mice, which have only initial signs of aging, e.g. at 6 months. Together with detailed analysis of axonal pathologies, microglia activation and behavioral tests this could elucidate some functions of Tspan2 in PLP-deficient mice.

ACKNOWLEDGEMENTS

I wish to thank Dr. Hauke Werner for offering me an interesting research project, scientific discussions, constant support and guidance during the rotation. I would like to thank Julia Patzig for valuable advices in data analysis, to Torben Ruhwedel and Dr. Wiebke Möbius for great explanation of technical questions and help with electron microscopy.

REFERENCES

Barry DM, Stevenson W, Bober BG, Wiese PJ, Dale JM, Barry GS, Byers NS, Strope JD, Chang R, Schulz JD, Shah S, Calcutt NA, Gebremichael Y, Garcia ML (2012) Expansion of neurofilament medium C terminus increases axonal diameter independent of increases in conduction velocity or myelin thickness. J Neurosci 32(18):6209–6219

Birling MC, Tait S, Hardy RJ, Brophy PJ (1999) A novel rat tetraspan protein in cells of the oligodendrocyte lineage. J Neurochem 73: 2600-2608.

Bjartmar C, Yin X, Trap BD (1999) Axonal pathology in myelin disorders. J Neurocytology 28: 383-395.

Bosse F, Hasse B, Pippirs U, Greiner-Peter R, Müller HW (2003) Proteolipid plasmolipin: localization in polarized cells, regulated expression and lipid raft association in CNS and PNS myelin. J Neurochem 86: 508-518.

Charrin S, Manie S, Thiele C, Billard M, Gerlier D, Boucheix C, Rubistein E (2003) A physical and functional link between cholesterol and tetraspanins. Eur J Immunol 33:2479-2489.

Chomniak T, Hu B (2009) What is the optimal value of the g-ratio for myelinated fibers in the rat CNS? A Theoretical Approach. PLoS ONE 4(11).

Edgar JM, McLaughlin M, Yool D, Zhang SC, Fowler JH, Montague P, Barrie JA, McCulloch MC, Duncan ID, Garbern J, Nave KA, Griffiths IR (2004) Oligodendroglial modulation of fast axonal transport in a mouse model of hereditary spastic paraplegia. J Cell Biol 166 (1): 121–131.

Gielen E, Baron W, Vandeven M, Steels P, Hoekstra D, Amerloot M (2006) Rafts in oligodendrocytes: evidence and structure-function relationship. Glia 54:499-512.

Griffiths I, Klugmann M, Anderson T, Yool D, Thomson C, Schwab MH, Schneider A, Zimmermann F, McCulloch M, Nadon N, Nave KA (1998) Axonal swellings and degeneration in mice lacking the major proteolipid of myelin. Science 280: 1610-1613.

Klugmann M, Schwab MH, Puhlhofer A, Schneider A, Zimmerman F, Griffiths IR, Nave KA (1997) Assembly of CNS myelin in the absence of proteolipid protein. Neuron 18: 59-70.

Möbius W, Patzig J, Nave KA, Werner HB (2009) Phylogeny of proteolipid proteins: divergence, constrains, and the evolution of novel functions in myelination and neuroprotection. Neuron Glia Biol 4: 111-127.

Nave KA (2010) Myelination and the trophic support of long axons. Nature reviews: Neuroscience 11: 275-283.

Sanchez,I, Hassinger L, Paskevich PA, Shine HD, Nixon RA (1996) Oligodendroglia regulate the regional expansion of axon caliber and local accumulation of neurofilaments during development independently of myelin formation. J Neurosci 16(16):5095–5105

de Waegh SM, Lee VMY, Brady ST (1992) Local modulation of neurofilament phosphorylation, axonal caliber, and slow axonal transport by myelinating schwann cells. Cell 68: 451-463.

Werner HB, Kuhlmann K, Shen S, Uecker M, Schardt A, Dimova K, Orfaniotou F, Dhaunchak A, Brinkmann BG, Möbius W, Guarente L, Casaccia-Bonnefil P, Jahn O, Nave KA(2007) Proteolipid protein is required for transport of sirtuin 2 into CNS myelin. J Neurosci 27(29):7717-7730.

Werner HB, Krämer-Albers EM, Strenzke N, Saher G, Tenzer S, Ohno-Iwashita Y, Monasterio-Schrader P, Möbius W, Moser T, Griffiths IR, Nave KA (submitted) A novel role of oligodendroglial tetraspans PLP and M6B in cholesterol trafficking and myelination.

Yool DA, Klugmann M, McLaughlin M, Vouyioklis DA, Dimou L, Barrie JA, McCulloch MC, Nave KA, Griffiths IR (2001) Myelin proteolipid proteins promote the interaction of oligodendrocytes and axons. J Neurosci Res 63:151-164.